Samsung Galaxy Tab 4 User Manual: Tips & Tricks Guide for Your Tablet!

By Shelby Johnson

www.techmediasource.com

Disclaimer:

This eBook is an unofficial guide for using the Samsung Galaxy Tab 4 tablet series and is not meant to replace any official documentation that came with the device. The information in this guide is meant as recommendations and suggestions, but the author bears no responsibility for any issues arising from improper use of the Galaxy Tab 4. The owner of the device is responsible for taking all necessary precautions and measures with the tablet.

Samsung Galaxy Tab 4, Samsung, and Galaxy Tab are trademarks of Samsung or its affiliates. All other trademarks are the property of their respective owners. The author and publishers of this book are not associated with any product or vendor mentioned in this book. Any Samsung Galaxy Tab 4 screenshots are meant for educational purposes only.

Contents

—

—

Introduction

The Galaxy Tab 4 series of tablets by Samsung were released starting in April 2014 as the successor to the popular Galaxy Tab 3 line of tablets. With this newly designed and updated device, owners are able to surf the Internet and stay on top of important tasks each day. Productivity and entertainment are taken to a whole new level with all of the new specs and features this latest device offers.

There are three different sizes of the Tab 4, with a 7-inch, 8-inch and 10.1-inch model among those offered. They each offer similar features, but the specs differ. For example, the 7-inch models (available in black or white), have a 1.2 GHz "tablet processor," while the 8-inch and 10.1-inch models have Snapdragon S5 1.2 GHz processors. In addition, the two larger models offer 16 GB of internal storage, and the ability to add as much as 64 GB more through an external microSD memory slot. The 7-inch models only have 8 GB storage and allow for adding just 32 GB more of external memory with a microSD slot.

All of the new Galaxy Tab 4 tablets feature the Android 4.4 mobile operating system, also known as "KitKat." They also offer decent on board memory with the 7-inch model sporting 1.5 GB DRAM, while the 8 and 10.1-inch models have 1 GB DDR3 RAM in them.

The Tab 4 also includes both rear and front-facing cameras that offer 3-megapixel and 1.3-megapixel shooting, respectively. The cameras allow for the popular mobile tasks of shooting photos, taking videos or having video chats with family, friends and colleagues.

Owners can opt for these tablets as Wi-Fi only models, or there are various service providers out there for 3G/4G LTE data plans. This will allow for wireless service pretty much around the clock for those who are looking to use the tablet quite a bit while on the go.

At first glance, this is a nice sleek tablet offered in black or white that should please Samsung enthusiasts as it rivals other competitors' tablets with its updated features.

What's New with the Samsung Galaxy Tab 4 vs. Tab 3?

Like the Galaxy Tab 3, the Galaxy Tab 4 comes in 3 different sizes. The Galaxy Tab 4 comes in 7, 8, and 10-inch vThe tablets are thinner and lighter making them easier to carry around and hold with one hand.

The latest Samsung Galaxy Tabs also have refreshed software, features, and interfaces. The overall design is different as well with the Tab 4s having a thinner bezel as well.

Specs for 3 Tablets

Galaxy Tab 4 7-inch Tablet

- 7.4 x 4.25 x 0.4- inches
- 9.6 ounces
- 1.2 GHz processor
- 1.5 GB of RAM
- 8 GB storage capacity
- 1.3 MP front camera
- 3.0 MP rear camera
- Android 4.4 KitKat

Galaxy Tab 4 8-inch Tablet

- 8.3 x 4.9 x 0.3-inches
- 11 ounces
- 1.2 GHz processor
- 1.5 GB of RAM
- 16 GB storage capacity
- 1.3 MP front camera
- 3.0 MP rear camera
- Android 4.4 KitKat

Galaxy Tab 4 10.1-inch Tablet

- 9.6 x 6.9 x 0.3-inches
- 1.1 pounds
- 1.2 GHz processor
- 1.5 GB of RAM
- 16 GB storage capacity
- 1.3 MP front camera
- 3.0 MP rear camera
- Android 4.4 KitKat

What's in the Samsung Galaxy Tab 4 Box?

Inside the box for the Tab 4, you'll find just a few other items. They include:

- Samsung Galaxy Tab 4
- microUSB to USB cable
- USB wall adaptor (for charging)
- Quick Start Guide
- Safety, Health and Warranty guide
- Paperwork related to Samsung, or any perks

These are all the essentials you need to get started with your tablet.

Setting up the Samsung Galaxy Tab 4

Once you have opened your tablet and all included accessories, connect the microUSB cable to the Tab 4 and the other USB end to your wall charger. It's best to plug this in and let the battery charge to full capacity, as the tablet may have shipped at less than half capacity.

After charging the device, power on your Galaxy Tab 4 using the power on/off button, located on the side of your tablet. The Welcome screen appears where you will select your preferred language. At this screen you can also choose to modify any accessibility settings, including those for screen timeout, spoken password characters, Talkback services, as well as Vision, Hearing, and Dexterity settings. It's important to take a look at these options from the start, as some settings may need to be adjusted based on your preferences. One example is extending the screen timeout, or time before your display goes dark, to a longer amount of time than the standard 30 seconds. Some of the other options are:

- Speak passwords – The device can be set to speak out each characters entered in password fields.
- Services – Talkback is OFF
- Vision – Font size is set at small
- Magnification gestures
- Negative Colors
- Hearing (sound balance, turn off all sound, Google or Samsung subtitles, etc.)

Note: *You can always go back to your Accessibility settings by dragging down from the top of your tablet's screen, and tapping on the "gear" icon for settings. Then Tap on the "Device" tab at the top menu, and scroll down to tap on the "Accessibility" option on the left side of the screen.*

Connecting to Wi-Fi

After adjusting any accessibility settings (you can also save doing that for later), you'll need to connect to your desired Wireless network (or 3G/4G service). If you're at home, an office or public place, you'll likely see the name of the appropriate Wi-Fi network listed. You can also press the "Scan" button towards the bottom of this screen to find available networks. Tap on the desired network to connect to Wi-Fi Internet service. Make sure to have any login password or other information you might need to connect with. Once connected, you'll see "Connected" under the network choice. You can tap on "Next" to move forward.

Next, you'll set up your date and time, as well as time zone. This may already have adjusted itself based on your location or Internet connection, but if net you can tap on the date, time or time zone to adjust it accordingly.

A EULA & Diagnostic Data agreement will pop up next. Read through this to decide if you want to allow it or not. Click on the "I understand" box and then either "Yes" or "No Thanks." Tap on Next to move to the next part of setup.

Adding a Google Account

A "Got Google?" screen will show up next, where you can answer "Yes" or "No" depending on if you use Gmail and Google apps. If you do, it's a good idea to integrate the various aspects of your account here, but you are not required to. You can always set this up at another time too. If you select "Yes" sign in to your Google account using your Email and Password. Click "OK" on the Terms of Service/Privacy Policy that pops up, after carefully reading the linked agreements. You'll be signed into your Google account, and a new screen will explain what Google services you'll be provided with while using the Tab 4. Tap on the small "Play" triangle in the lower right corner to move forward.

Next you'll get a "This tablet belongs to..." screen. Your name may be automatically inserted into the fields already, or you can choose to enter a name or modify the name shown. Hit "Enter" and "Done" to move forward. Your Google account will be connected with your device.

Create a Samsung Account

After this step, you're given the option for creating a Samsung account. You can either sign in if you've already got an account, or go through the steps to create a brand new account. A Samsung account comes with a variety of features including the ability to download Samsung GALAXY Apps, use ChatON (a global messenger), play Samsung games or videos, use Samsung Books, use various Health tracking features, and Manage the contents of your device anywhere, among other helpful options.

Creating the account will require entering your email, password, date of birth, name, and zip code. You'll also need to verify the account through an email they will send to the email account you use to sign up with. The Samsung account is free to sign up for, but not required to use the tablet. You can also choose to create a Samsung account at another time.

Final Setup Steps

After moving past the Samsung account step, you may be presented with free Dropbox storage (as part of a perk). With the tablet you can choose to sign up for this free online storage service, or you can skip this particular step as well.

The final step of the initial setup will involve entering an identifiable name for your Galaxy Tab 4, so you can easily recognize it via Bluetooth, Wi-Fi or other devices you might connect to it with.

Setting up your Email, Facebook and Twitter Accounts

When you are ready to set up your email accounts, whether it is your Gmail, Outlook, Yahoo!, or a combination of Hotmail and AOL, you can access this function easily to ensure all of your communications come directly to your tablet.

1. From your Home Screen, tap the "Apps" icon.
2. Tap the "Email" icon.
3. Enter the Email address you wish to add.
4. Enter your password.
5. Tap Next.

Your tablet and network will detect your settings. The only thing you have to do is set the frequency you would like the device to retrieve your email messages. You can also set notifications, so you are aware you have a new email. Name the email, whether it is work, personal or otherwise, and tap done.

Facebook

Equally as simple as the email setup, simply follow these easy steps:

1. From the Home Screen, tap the Google Play Store icon.
2. Search "Facebook."
3. Tap the App for download.
4. Wait for the App to download completely.
5. Tap to open, and enter your Facebook username and password.

Twitter

No tablet is complete with the Twitter app, and the TAB 4 is no exception! To set up your account, simply follow these quick instructions:

6. From the Home Screen, tap the Google Play Store icon.
7. Search "Twitter."
8. Tap the App for download.
9. Wait for the App to download completely.
10. Tap to open, and enter your Twitter username and password.

Start Tweeting to your heart's content!

Contacts

You can save your contacts on your Galaxy Tab 4, so that you will be able to quickly access information about your family, friends, and colleagues and send them a message.

Adding Contacts

1. Touch Contacts and select Create.
2. Touch each field to enter the contact's information.
   - **Picture:** Touch the picture icon to assign a picture to the new contact.
   - **Name:** Enter contact's name. Touch to display additional name fields.
   - **Phone:** Enter contact's phone number.
   - **Email:** Enter contact's email address.
   - **Groups:** Assign the contact to a group if needed.
   - **Add another field:** Add additional fields for the contact if necessary.

3. Touch Save.

Linking Contacts

The Tab 4 can synchronize contacts among different accounts like Google, Corporate Exchange, and many other email providers. To link contacts, do the following:

1. Touch Contacts.
2. Touch a contact in the Contacts list to display it.
3. Touch Menu and select Link contact.
4. Touch the contact you want to link.

Simple repeat steps 3 and 4 to link other contacts.

Note: *Linking contacts is helpful because any time you change the contact's information, it will update across your accounts.*

Sharing Contacts

You can send a contact's information using the Tab 4's Bluetooth capability to other Bluetooth devices, or in an email or as an attachment.

1. Touch Contacts.
2. Touch Menu and select Share.
3. Touch selected contact to share only the contact displayed, or touch multiple contacts.
4. Touch a sending method like Bluetooth or email.
5. Follow the prompts to share the contact.

Note: *Not all Bluetooth devices accept shared contacts and not all devices support transfers of multiple contacts. To find out if yours does, simply check the target device's documentation.*

Setting Up Security on the Galaxy Tab 4

With the Galaxy Tab 4, you can use the Security settings to secure your device by setting up a passcode to access the home screen, or going a step farther and encrypting the information on your SD Card. To do so, tap the "Settings" icon, select "General" and then "Security."

To create a passcode, simply turn on the passcode slider, and select a four-digit code that must be entered by the person accessing the tablet's home screen. You can set the duration you prefer before the passcode request is shown, as well.

To apply encryption security to the device, select "Encrypt Device". The displayed screen that follows will walk you through the process completely. You can encrypt the external SD card data, which will require a password each time it is inserted for access. This encryption process will take approximately one hour, so be sure your device is powered to at least 80%, or is plugged in during the process.

Customizing the Home Screens

The Home Screen is where it all begins on your device. It is the starting point for every movement, and can be customized to fit your personality or use rather quickly, so your tablet is a reflection of you, your organization and your needs as a whole.

First, to change the wallpaper on your screen, to reflect your interests and not those of Samsung's factory application, take these quick and easy steps to beginning your tablet personalization.

Changing Wallpaper

1. Touch & Hold any Empty Area on the Home Screen
2. Tap Set Wallpaper
3. Choose Home Screen, Lock Screen, or Home & Lock Screens
4. Select the Image you Want to Appear from: Gallery (Your Images Saved on the Device); Live Wallpapers (Animated Options, Not Available for Lock Screen); Wallpapers: (Preloaded Wallpaper Options)
5. Tap Done or Set Wallpaper

Adding & Deleting Shortcuts

Adding shortcuts to your home screen will allow you to access all of your favorite apps quickly and effortlessly, so you do not waste time in getting the information you want or need from your tablet. This process allows you to fully customize your home screen.

1. Choose a Home Screen to Add a Shortcut to, and Tap "Apps"
2. Locate the App you Desire, and Touch and Hold its Icon (A shortcut will be created, and the app menu will close)
3. Hold the New Icon (it will appear hovering over the screen) & Position it on the Screen and Release

Adding a Shortcut

1. Choose a Home Screen and Touch & Hold an Empty Pace
2. Tap Apps and Widgets
3. Touch & Hold the Icon of Choice
4. Position the Icon on the Screen & Release

Deleting a Shortcut

1. Touch & Hold the Shortcut you want to Delete (Edit Screen Will be Activated)
2. Drag the Icon to the Remove Bar at the Top of the Screen & Release

19

Adding a Widget

1. Tap the Apps Icon, then the Widgets Icon
2. Select to the Widget you Want to Apply to the Home Screen
3. Touch & Hold the Widget Until the Edit Screen Unveils
4. Drag the Widget to the Screen you Wish it to Appear
5. Release the Widget in Place

Removing a Widget

1. Touch & Hold the Widget you want to Delete (Edit Screen Will be Activated)
2. Drag the Icon to the Remove Bar at the Top of the Screen & Release

Creating & Managing Folders

For organizational purposes, folders can be created and placed on the Home Screen, and application shortcuts can be placed within to keep your screens tidy. Here's how:

1. Navigate to the Home Screen Where You Would Like the Folder to Appear
2. Touch & Hold an Empty Area and Tap Folder
3. Enter a Folder Name and Tap OK
4. An Empty Folder is Applied

When you add applications to the home screens, they can be dragged into the folders as you see fit. This allows you to separate work from play, or even designate folders for different users if necessary.

Samsung Galaxy Tab 4 Basics

Learning your way around the Samsung Galaxy Tab 4 is an intuitive process, which means your brain will begin showing you how it works in no time. It never hurts to know what you are looking at to begin with, however, to get you heading in the right direction!

Status Bar

In some cases, the status bar may be hidden, so to display it simply touch and drag the top of the screen down. It will then disappear a few seconds later, but can be accessed at any time.

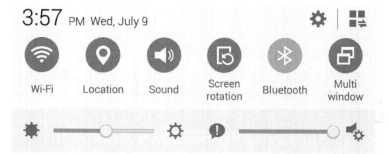

Notifications Panel

At the top of the screen, notifications icons will appear on the Status Bar to alert the user to scheduled events on their calendar, device updates and much, much more. To access these notifications, simply drag the status bar down, and open the Notifications Panel.

Apps

The Apps Screen allows you to access all of your apps and widgets with a tap, and is located in the lower, right-hand corner of the home screen. It appears as a square box, composed of dotted lines. From this area, App Shortcuts can be created, as outlined previously, so they appear on the home screen for easy access to all of your favorite games, informative topics or even banking.

Settings Screen

The Settings Screen is an important part of how your Galaxy tablet operates. This segment allows you to configure your device, set application options and add accounts to help personalize your experience.

To access the Settings Screen, touch "Settings" from the Home Screen. Also, from any screen you can swipe downward to display the Notification Panel and then tap "Settings."

Settings Tabs

The Settings are separated into four main groups or topics. Once you access Settings, the following groups will appear:

- Connections: Controls your Device's Wireless Connections
- Device: Personalizes your Sounds, Displays, Accessibility & Input
- Controls: Configures Language, Input, Motions & Smart Screen
- General: Create & Modify Accounts (Email, Google, Samsung, etc.); Manage Security, Locations, Storage and Other Device Features

The Settings Screen

At first glance, users will notice that there are a number of ways that options are enabled or disabled, including:

- On/Off Sliders
- Check Mark Displays

Use the appropriate option where applicable.

Connections

The Connections segment controls your device's wireless connection, so you can connect to the Internet where a Wi-Fi capable connection exists. Simply choose the Wi-Fi option available when this segment is displayed and enter the password when prompted.

This segment also allows users to connect to Bluetooth devices that are within 30 feet of the device, including wireless speakers, keyboards, and headsets. Simply turn the Bluetooth capability on, and tap the device that is located for use to move forward with the connection.

Airplane Mode

Airplane mode allows users to enjoy their tablet's features including games, music, videos or cameras while on an airplane or any other area where accessing a network is prohibited. This means you can still enjoy your tablet, without interfering with the mechanical or electronic signals around you. You will NOT be able to access online information or use certain application as a result of Airplane Mode being enabled.

Location Services

Location services allow you to find places near you, using Google Map, or to scout out restaurants nearby that suit your taste at any particular moment. You must enable location services on your device for certain apps, including GPS or mapping apps to work properly. At times, you may be prompted with a message that says, "X-App wants to access your current location" and by hitting yes, the results you receive will be more accurate.

Device

This section of the Settings screen contains settings for your device, including sounds, volume and notifications.

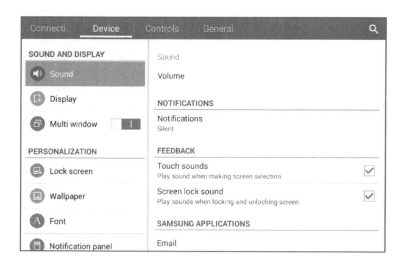

Sound & Volume

You can set the volume level for a bevy of sounds that emanate from your tablet by simply tapping "Sound" from the settings screen, and dragging the sliders to set the volume to your liking for:

- Music, Video, Games & Other Media: Sounds made by music apps, video apps, and more.
- Notifications: Sounds for alerts to new messages and other events.
- System: Application sounds and other sounds made by your device.

Notifications

To create notifications that alert you to the things that are important to you like email, instant messages, social media updates, etc., you can designate a different sound for each, so you know what to expect when you hear the notification. To select a default sound for messages, alarms and other notifications, simply:

Start from the Settings screen, tap Device, tap Sound, and finally Notifications. You can sample the sounds before selecting the option for different alerts.

Display

Use the Display settings to configure the way your device's screen operates by beginning from the Settings screen, touching Device and Display.

The following options are available:

- Brightness: Touch and drag the slider to set the brightness.
- Screen timeout: Set the length of delay between the last key press or screen touch and the automatic screen timeout (dim and lock).
- Daydream: Set the device to launch a screen saver when your device is connected to a desktop dock or charging. Touch the OFF/ON button to turn Daydream ON . Touch Daydream for additional options.
- Show battery percentage: Set the device to display the remaining battery life.

Screens

When it comes to your home screens, you will notice that the device comes with five as a default, but you can add up to seven, should you desire. This will provide plenty of room for apps, so you can organize your screens as you see fit.

Adding and Deleting Home Screens

You can arrange the screens in any order you choose, simply follow these directions!

- From any Home Screen Make a "Pinch" Motion using Two Fingers to Display the Edit Options
- Touch & Hold the Home Screen you Would Like to Move
- Drag the Screen to the New Location & Release.

Keep in mind, when you move a Home screen, the other Home screens will be reordered automatically. Simply press the Home button to return to the main Home screen.

Add a Home Screen

- Touch & Hold an Empty Area on the Home Screen & Tap Page
- Press the Home Button to Return to the Main Home Screen

Deleting Home Screens

You can have from one to seven Home screens on your device. They can be deleted or added at any time.

- From any Home Screen Make a "Pinch" Motion using Two Fingers to Display the Edit Options
- Touch & Hold the Home Screen you Would Like to Delete
- Drag the Screen to the Remove Bar at the Top of the Screen
- Press the Home Button to Return to the Main Home Screen

Galaxy Tab 4 Features

The Galaxy Tab 4 has some great features that make this tablet unique from the rest. The following sections describe some of the tablet's features.

Nearby Devices

Nearby devices allows you to share content with other Samsung, NFC or WIFI Direct users. This is the cool tapping to share function you see on the commercials. To enable the functionality, follow these quick steps:

1. Select the "Apps "icon in the lower, left-hand corner of the device.
2. Select Settings.
3. Select Connections.
4. Select Nearby devices and slide to right.
5. Follow the instructions to allow your compatible devices to share information.

The device you are sharing with must follow the same steps! Once they are both activated, simply locate the content you would like to transfer, whether it is an image, music, links or otherwise. Once you have located the content, place the two devices together back to back until the message "Touch to beam" appears. Tap the message and await the prompt to separate the devices. Once you have, the transfer will begin. Once complete, the second device will show the content on its screen.

S Voice

Similar to the iPhone's Siri, Samsung has developed the S Voice for many of their devices. In order to activate the function, press the home screen button twice. You can pick from any of the boxed commands, like setting a timer or getting an answer, or you can command your device to look up something online, find a Chinese restaurant or play something from your playlist. It will also give you turn-by-turn driving directions, just like a GPS device would.

To use complete, tap your "Home" button twice, or use the following steps:

1. Click on the "Apps" icon from the bottom menu.
2. Click on "S Voice."

Begin giving commands to your Samsung Galaxy Tab 4.

Motion Control

For those who prefer not to touch their device to find what they need, say if your hands are covered in dough or cake batter, and you really need to find the next step of a recipe, there is a Palm motion feature that allows you to operate the tablet without actually touching it.

To activate this feature from the Apps screen:

1. Tap Settings.
2. Tap Controls.
3. Tap Palm motion, dragging the Switch to the right to activate.

Once activated, you will be able to swipe your hand from left to right or up and down, depending on how the content is displayed to move forward or backward with a motion from your hand.

Also in the Palm motions category is Mute/pause, which allows you to cover the screen with your palm to mute incoming calls, alarms or music that is playing. Finally, you can pause the content onscreen simply by looking away – which is perfect for watching videos, should you become distracted. In addition, you can record a screen capture – after ensuring the feature is turned on in motions and gestures – simply by swiping your hand from one side of the device to the other. The image will appear in your camera roll once taken. This option is not available within every app.

Multiple Users

The Samsung Galaxy Tab 4 has the ability to support multiple users, which means you can share the device with your roommate, friend, partner and even your kids! All you have to do is create user accounts and profiles for each person using the tablet, and the available content will become the default applications.

Each user can customize the tablet to fit their profiles, including their own apps, wallpaper and tablet settings that reflect their personality.

Setting Up Users and Profiles

To set up different users and profiles, simply:

- • Begin from the Settings Screen and Tap General, then Users
- • Touch Add User and Select "User" or "Restricted Profile" (Restricted profiles are restricted accounts that allow only limited access to apps and content.)

- • Follow the Onscreen Prompts to Proceed with the Account Set Up

Once multiple accounts are set up, profiles will appear at the top right hand side of the locked screen.

Changing Users

To begin bouncing between users, select the profile at the top that belongs to you (or to the other person, if he or she is using the device.

Change User Account Nicknames and Photo IDs

The photo and name each user uses in their personal Contact entry (listed under ME at the top of their Contacts list) is used as their User account ID. To change a user's photo or name:

- • Tap Contacts
- • Tap their Personal Contact Entry to Access Edit

Deleting a User or Profile

Each user is capable of deleting their personal account, but only the owner -- the person who has registered the tablet -- can delete the owner account, and any other if they wish.

As the Owner:

- • Begin at the Settings Screen, Tap General, and Users
- • Tap a User Profile & Touch Delete

If you are a User:

- • Begin at the Settings Screen, Tap General, and Users

- • Touch Menu & Delete [user] from this Device

Kids Mode

Kids Mode allows parents to set limits for their children, restricting accessible apps and managing the length of time they spend using the device. The app will need to be downloaded for the tablet, and once it has completed, usage is simple.

- Tap the Kids Mode Icon
- Set a PIN for Safety
- Set a Profile (Kid's Name, Birthday, Image, etc.)
- Each App Name on the Tablet will Appear; Check Those the Child is Cleared to Use

There are only two screens available in Kids Mode, so it is simple to use and access the approved apps. In addition, Kids Mode comes preloaded with music, games and a camera that allows the child to take pictures as he or she sees fit, just like the adults!

Smart Stay

Smart Stay is not new to Samsung devices, but it has been reported that it works better with this one. This function allows the screen to stay active as long as you are looking at it.

This means it will not dim or return to idle. This is great when you are watching a video, as it pauses the video when you look away.

To turn Smart Stay on, complete the following steps:

1. Select Settings.

2. Select Controls
3. Select Smart Stay.

Alternatively, you can swipe down from the top of your screen for the "Notifications" menu, and swipe across the top until you see the "Smart Stay" icon and tap that, making it turn green.

Smart Pause

Smart Pause allows you to watch a video and pause it simply by looking away! If you have ever started a video, only to be distracted by someone or something, and miss it in its entirety, Smart Pause is exactly what you were missing! Smart Pause will pause the video when you move your eyes away from the screen. To enable this feature, simply go to the setting page, display additional options, and slide the button for "Smart Pause" to green.

Toolbox

Toolbox is a floating app drawer that allows you to access the apps you use most, so they are always at your fingertips without sliding around your home pages. This toolbox holds five apps, which will be automatically added when the option is enabled. No worries, however, you can change them.

To activate the toolbox, simply pull down the screen from the top with two fingers and locate the Toolbox icon. Tap the icon to turn it on. To move forward, from the settings menu, scroll down until you see "toolbox" Tap the icon to be transferred to the Control Panel where you can choose which apps you wish to appear in the toolbox, by toggling them on (or off, should you want to change the lineup at any time). Hit save, and your toolbox is ready for use!

Using Samsung Galaxy Tab Modes

There are several different modes you can use on the Galaxy Tab 4. From Airplane Mode to Easy Mode, there are many different ways to use the modes on this device. The following sections discuss the modes and how to use them when you need them.

Using U. Power Saving

The amount of remaining battery power on your tablet only becomes an issue when you are lacking battery power for it. Fully charged devices are rarely fretted over, especially when you are near a power supply. When you are not, and find yourself low on power percentages, there are a number of ways to save what you do have – in case of an emergency.

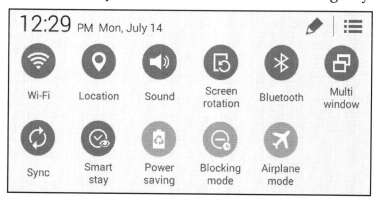

The goal is to save battery power by limiting the device's functions, and to do so you simply have to follow these steps:

Swipe down from the top of your screen to bring the Notifications/Quick Settings panel.

1. Tap the far right-hand corner to expand the quick settings.
2. Tap Power Saving.

Select from the following options to enable power saving:

- **Block Background Data:** This prevents apps that are running in the background from using mobile data connections.
- **Restrict Performance:** Limits various options, including backlights that cause the battery to drain.
- **Grayscale Mode:** Displays grey tones, instead of color, to limit the battery drain from displaying color.

Using Screen Mirroring

Screen Mirroring allows you to stream various content from your Samsung Galaxy Tab to other compatible devices wirelessly, like your television or another compatible device. You can also purchase an AllShare Cast Wireless Hub, a device that connects to a TV monitor and will allow it to work with Screen Mirroring. The feature works only on the Samsung Galaxy Tab 4 8" and 10.1 models as of this publication's first edition.

If you already have a Samsung TV monitor, it may include Screen Mirror (or casting) as an option. To perform screen mirroring with a compatible TV or AllShare Cast Wireless Hub:

1. Use your TV remote, select the "Input" button (or on-screen option) and then select "screen mirroring" (or similar option).
2. On your Tab 4 tablet, swipe down from top of screen to bring down Notifications/Settings.
3. Go to your "Settings" and then to the "Network Connections" tab.

4. For Screen Mirroring, select the compatible TV or the AllShare Cast and move the slider to "ON" position to begin screen mirroring. You should see everything shown on your tablet on the TV screen now.

Note: *To shut off the feature, you will need to go through similar steps as above and move the slider to off. Turning your tablet or TV off will likely end the mirroring as well.*

This feature allows you to share videos, pictures and content with a group, without huddling around your device.

Using Multi Window

For all of the multi-tasking people out there, the Galaxy Tab 4 does not disappoint, allowing users to have several apps open at once, signaled by their availability to swap back and forth by their inclusion in a tray that exists on either side of the tablet's screen. The following image displays an example of the multi-windows, with the left window on top of the main apps screen.

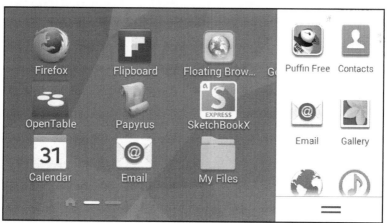

Here's how to enable it:

1. Open the "Notifications" screen by swiping down the screen with from top to bottom.
2. Tap "Multi-Window."
3. A Tab will appear on the left of your screen.
   Note: *To drag the tab from one side of the screen to the other, hold the tab down, and move it to the other side of the screen.*
4. Tap "Edit" at the bottom of the tray to swap apps.
5. Simply drag the available/compatible apps that appear during editing to the tray.
6. Open/Close the tab as you see fit for a fast approach to using your apps.

This means you can add the apps that you want, like email or Internet, and get to them easily without flipping through your screens. Obviously this is helpful with apps you use the most often, and it can really save time if you keep them well organized.

Using Airplane Mode

It is possible to use your Galaxy Tab 4 on the airplane, so you can enjoy games, or even schedule calendar events while you travel.

This disables your tablet's ability to transmit a signal that would usually interfere with the plane's functionality while allowing you to still use it. Here's how:

1. Swipe down from the top of screen.
2. Tap on the "Airplane mode" icon.
3. Click "OK" on the pop-up box that gives you details about the mode.

An alternate way to activate Airplane mode:

1. Tap on your "Apps" icon in bottom right corner of screen.
2. Select Settings icon.

3. Select Connections tab.
4. Side Airplane Mode to the right.
5. Click "OK" on the pop-up box that gives you details about the mode.

Airplane mode is now enabled. To disable Airplane mode, simply repeat above instructions.

Alternatively, you can swipe down from the top of your screen for the "Notifications" menu, and swipe across the top until you see the "Airplane mode" icon and tap that, making it turn green.

There are other times when Airplane mode comes in handy when you do not want notifications to come through to your tablet. For instance, if you are using your tablet over a speaker system to play music, this is a good time to use Airplane mode. Another time may be if you are using your tablet to play a video for a group of people.

Using Private Mode

Private Mode is an option that allows the users to hide images, videos and content from the view of others. Users will be able to view all of the content on their device, but when another person accesses it, the files contained within will not be visible. To view the private content, a passcode or pass pattern is required.

To Enable Private Mode

1. Swipe down from the top of any screen.
2. Tap Private Mode from the list of icons.

There is a short tutorial that will users will walk through, before being asked to enter a pin code. This will only happen the first time.

When you are in private mode, you are the only person who can access the hidden files, so be sure to disable it should you fear someone else flipping through your personal content. To do so:

1. Swipe down from the top of any screen.
2. Tap Private Mode from the list of icons.

This will allow your device to revert back to "normal" with your hidden files securely placed out of reach.

Adding Files to Private Mode

Currently, videos and images can be saved in private mode. To place them there:

1. Turn Private Mode On
2. Navigate to the Photo or File you want to Appear in Private Mode Only
3. Select One or Multiple Files and Tap the Overflow Menu Button in the Upper Right of the Screen
4. Tap on Move to Private.

Using Blocking Mode

Blocking Mode allows users to turn off notifications, alarms, LED indicators, and more simply by enabling the feature. This is great during parties or meetings, so your tablet isn't disturbing anyone, but is still receiving information in real time. To enable it:

1. Begin at the Settings Screen, tap Device, and Blocking Mode.
2. Tap the OFF/ON button to turn Blocking Mode ON (or OFF, when applicable).

From this area you will be able to decide which options you turn off (or back on), including notifications and alarms, and also allows you to choose a time range you would like the blocking to occur.

Samsung Galaxy TAB 4 Apps

Although different people enjoy different things, no matter where you land on the scale of conservative to liberal and beyond, there is certainly an app that will suit your fancy. With 1,295,454 Android apps on the market at last count, there is something for everyone. When it comes to music, reading and video, some of the best are available for all entertainment needs.

Music Apps

Music is a large part of any electronic device, and when you can hear your entire favorite tunes in one place, it can change the outlook of your day.

Milk Music

Milk Music is similar to Spotify or Pandora, but does not require payment or advertising supported listening to allow you to enjoy all of your favorite music. Simply download the app from the Samsung app store, as this particular jukebox-like entertainment is free with your Samsung purchase.

Reading Apps

If you are traveling with your tablet, there is a pretty good chance you are going to want to have a book or two on hand even if they are being read during a morning commute on the train. If so, look for two of the best in Kobo and Kindle. Both are available on the Google Play app store, and since they compete with each other, their offerings and service simply keep climbing to outdo the other!

Video Apps

When it comes to video apps, most aren't too picky about what they choose -- as long as it plays. However, MoboPlayer provides a little something extra in the form of stop and go video, which means it will pick up where you left off on your favorite video entertainment options.

Camera Apps

The camera on the Galaxy Tab is incredible, so taking perfect images will never an issue. However, should you want to apply filters, use timers and take the best picture you can with your device.

Camera ZOOM FX

This app allows you to set timers and use motions to take pictures, before balancing the colors in the edit section that result from the amazing "burst" shot option of taking up to 20 shots per second. Long exposures are available too, and all can be recorded using the customizable camera buttons that you can place where you would like (perfect for lefties). In addition, there is a stabilization meter, horizon level indicator, customizable grid overlays and voice and sound activated triggers. The price is $2.99 for the app, but well worth it.

If you are looking for something free, Camera360 Ultimate is full of features that allow you to apply filters and specify which type of picture you're shooting, to differentiate between people and plants. This app provides quick sharing and photo effects along with a seriously easy to use interface.

Samsung Galaxy TAB 4 Tips and Tricks

There are a number of ways that you can enjoy Your Samsung Galaxy Tablet for years to come, with ease and success. The following tips and tricks will get you through the technological ups and downs that come with the purchase of anything new!

How to Extend the Battery Life

Extending your battery life revolves around managing the battery sucking options that come at a higher default level than you require to enjoy your tablet. Here are a few tips:

- Adjust the automatic brightness level on the device.
- Use a Wi-Fi connection when one is available.
- Set your display timeout for as little as possible (minimum 15 seconds).
- Use Power Saving Mode.

These tips will keep you battery operating at optimum power, so you are never without a charge in an emergency situation. Each can be found under Settings, and manipulated to fit your power saving needs.

How to use Multi Window

Multi Window allows you to operate two apps at once. To do so:

1. Start from the Home screen, Tap Settings, and then Device.
2. Tap Multi Window (activate by sliding the tab to the right).

You can only use the apps that appear in the multi-window caddy in a split screen.

How to use Smart Remote

Smart Remote is an app you can download to your tablet, and use on the electronic devices around your home. Here's how:

1. Locate the Smart Remote icon and tap it.
2. Identify your location and service provider when prompted.
3. Tap the brand of your TV (tap show other brand to reveal more options).
4. Make sure the top of your device is pointed at the TV, unobstructed.
5. Follow the prompts and signals as the device finds the best connection.

You can now control the device you programmed directly from your tablet!

How to Share With Other Samsung Devices

Galaxy Tablet users can send information to other Samsung users effortlessly using their devices, without lifting a single pad or paper!

First, users can send their personal contact information to other Bluetooth devices:

1. Tap Contacts.
2. Tap your personal contact entry to display your information.
3. Tap Menu & tap share namecard via (choose method).

Sharing Others' Contact Information

1. Tap Contacts.
2. Tap Menu and tap Share namecard via (choose method).
3. Tap selected contact (or multiple).
4. Tap sending method.

Share Shot

To share a shot you have just taken with Samsung friends, the process is equally as easy! Tap Options and select Share Shot to share a photo directly to another device via Wi-Fi Direct.

Sharing Images

Sharing images that you have on your device is easy and fun, so others can see exactly what is happening from your vantage point!

Use one of the following methods to share images:

In a folder, touch images to select them, and then touch to send them to others or share them via social network services. When viewing an image, touch to send it to others or share it via social network services. Tap Menu, Tap "Share Via" and select videos by ticking or touching followed by "done" then choose the sharing method.

How to Listen to Music

Listening to music will require a music app or Google Play Music service on your device. The speakers are built into the device, and you can simply hit the volume button on the side to turn it up and down. Within the music app you can scroll via artists, playlists, tracks and albums, depending on your mood. You can also connect headphones or a Bluetooth headset to enjoy the music privately.

How to Play Videos

Videos will appear with an arrow on the lower left-hand corner denoting that they are actual videos, instead of stills. Tap the video and it will begin playing on the default video player if you have not download an optional version.

How to Watch TV

Movies and television are available for viewing instantly through streaming, or downloaded and viewed later through the Google Play store. Apps like Hulu Plus and Netflix will allow you to stream your favorites on demand. Simply tap your "Play Movies & TV" icon to begin.

WatchOn

WatchOn transforms your Samsung device into your home television by allowing you to connect to your service from anywhere, accessing your favorite shows in seconds. You can control can your TV, set-top box, DVR and other devices from your Samsung device and never miss a second of the shows you forgot to record while you were at home, or those you have been stockpiling for some time. You can even transfer the content to a screen bigger screen near you, say at a friend's house, instead of watching it from your device. It is important to program your WatchOn account from your home, with active Internet service. The prompts will put you through the motions without issue, and open up an entirely new world of entertainment.

How to Set Reminders & Alarms

Users can set a reminder for themselves to take medication, make a phone call or show up to a meeting a little early, simply by setting an alarm. Here's how:

Set alarms for reminders or for wake-up times.

1. Touch Alarm.
2. Touch Turn Alarm On to activate.
3. Change the timing, add the reminder note and hit save.

A clock will appear in the status bar when an alarm or reminder has been set.

How to Change Galaxy Tab 4's Wallpaper

To change the wallpaper on your screen to reflect your interests and not those of Samsung's factory application, take these quick and easy steps to beginning your tablet personalization.

Changing Wallpaper

1. Touch and hold any empty area on the home screen.
2. Tap Set Wallpaper.
3. Choose Home Screen, Lock Screen, or Home & Lock Screens.
4. Select the image you want to appear from: Gallery (your images saved on the device); Live Wallpapers (Animated Options, Not Available for Lock Screen); Wallpapers: (Preloaded Wallpaper Options).
5. Tap Done or Set Wallpapers.

How to Take Pictures

From Home Screen, tap the app icon and then the Camera icon.

1. Touch the image on the preview screen.
2. When the image you want is in focus, the focus frame turns green.
3. Touch the camera Icon to take the photo.

The images you take will be stored in your camera roll, called "Gallery" and can be retrieved or deleted at any time.

How to Take Video

Recording videos is fun, and really help you take the memories that are being made now, and storing them for a lifetime to come.

1. Touch the video camera icon to record video.
2. To pause recording, tap the pause icon.
3. To continue recording, tap the circular button.
4. To stop recording, tap the stop square.

How to Take a Screenshot on Galaxy Tab 4

Want to take a screenshot on your tablet? The process is relatively easy, should you want to show someone a particular screen on your device. Simply press and hold down the oval "home" button (bottom of device) while also pressing and holding down the power button (on side of device). You should hear a shutter click noise, indicating a screenshot was taken. You can find the screenshot by going to your "My Files" icon on the tablet and then into "Images."

How to Use Dropbox for File Transfers & Storage

Dropbox is an exceptional file storage and transfer option that comes standard with your Galaxy device. It allows you to share files effortlessly simply by dropping them into the folder on screen, and allowing others to retrieve them.

This cloud storage option places a large data storage opportunity at your fingertips, so you do not have to send bulky emails. When you save files to Dropbox, your device automatically syncs with the web server and any other computers that have Dropbox installed. Simply:

1. Tap the Dropbox icon.
2. Follow the onscreen instructions to complete the initial setup (you only have to do this once).

Once it is set up, you literally touch and hold the file you want to place in the Dropbox and drag it to the icon for storage. You can also review the content others have placed inside for your review directly from your device.

How to Use Google Drive

Google Drive requires a Google account to access and use, but once you have one you can begin accessing online documents including spreadsheets, presentations and even Word documents online. As a cloud service, Google Drive allows you to create a menagerie of documents and store them to share with others.

You can download the Google Drive app from the Google Play Store, and access all of your created or shared documents effortlessly. You can change the names, content and other information on the documents and re-share them with others or just store them for future use.

How to Pair Bluetooth Devices

Pairing Bluetooth devices allows you to use a wireless keyboard, speakers or headphones effortlessly to enjoy your device all that much more!

To sync your Bluetooth device to your Samsung:

From any screen, touch the top of the screen and swipe downward to display the Notification Panel. Tap Bluetooth to ensure it is turned on. Once Bluetooth is on, the device will begin scanning for the items nearby. Once it has located the one you wish to use, tap the device and allow it to pair up with the Samsung. Depending on your Wi-Fi signal, pairing can take time.

How to Print from Galaxy Tab 4

To manage your printers:

1. From a Home screen, touch Settings.
2. Choose Connections tab.
3. Choose More networks.
4. Touch Printing.
5. Once in Printing, choose a print service to configure the printer.

This will work if you have a Samsung printer.

Another way to print from the Galaxy Tab 4 is to use a Bluetooth printer. This process has several steps, but it is certainly achievable. To print do the following:

1. Pair your Galaxy Tab 4 and your Bluetooth printer.
2. View the document, webpage, or image you need to print.
3. Choose the Share button. If a Share button isn't visible simply press the Menu button and tap the Share button.
4. Choose Bluetooth from the Share.

5. Choose the Bluetooth printer from the items on the Bluetooth Device Picker.
6. If a prompt appears on the printer, confirm.
7. The document is uploaded, and then it prints.

If you have an HP ePrint enabled printer, you can also print from your Galaxy Tab 4. To print to an HP printer, do the following:

1. Download the HP ePrint app to your Galaxy Tab 4.
2. Use the app to print over a Wi-Fi network.

This app works by sending an email to your printer to print.

How to Backup and Restore

You can backup your tablet to the Google server. To enable or disable backup to the Google server do the following:

1. From a Home screen, touch Settings, choose General tab, then choose Backup and reset.
2. Touch Back up my data to enable or disable backup. Application data, Wi-Fi passwords, and other tablet settings will be backed up to the Google server.

NOTE: *Once the backup my data option is enabled, then the Backup account and the Automatic restore options are available for your Galaxy Tab 4 tablet.*

How to Enter Text by Voice

The Galaxy Tab 4 uses Google Voice Typing to enter text by voice. To enable Google Voice Typing, do the following:

1. From a Home screen, touch Settings, choose Controls tab, then choose Language and input.

2. Touch Google voice typing to automatically use Google voice typing.

3. Touch the settings button next to Google voice typing and choose from the following options:

- **Choose input languages:** Touch Automatic for the local language or select a language.

- **Block offensive words:** Here you can enable or disable blocking ofcertain offensive words.

- **Offline speech recognition:** Here you can enable voice input while offline.

How to Hard Reset Tablet

In some instances, a hard reset (also known as "factory reset") may be your final option for trying to fix an issue, or you may simply want to erase all the data on your tablet (in the case of selling a used tablet or giving it to someone).

```
Android system recovery <3e>
KOT49H.T230NUUEU0ANE2

Volume up/down to move highlight;
power button to select.



apply update from ADB
apply update from external storage
wipe data/factory reset
wipe cache partition
apply update from cache
```

Here's how to do a hard reset on the Tab 4:

1. First turn off the tablet.
2. Press and hold down the Home + Power button + Volume Up buttons all at once. After it has started up, you'll see a Samsung Tab screen and then you will see the screen in the image above these instructions.
3. Make several selections on these screens that come up. To navigate up and down on the menu choices, use the Volume buttons. To make a selection press the Power button.
4. Choose "Wipe data / factory reset."
5. Choose "Yes - erase all user data."
6. Choose "reboot system now."

These steps should factory reset your Tab 4 tablet to its original state. Remember it will erase all data on the device, so make sure you have backed up any important files that were only on the Tab 4.

Great Apps to Improve Your Samsung Galaxy Tab 4

The Galaxy Tab 4 comes with plenty of great apps already built in, including those for email, contacts, notes, calendar reminders and more. However, there are plenty of great free apps out there to install from the Google Play Store. Here's a look at 20 of the better apps you might want to consider for your device.

1. Facebook – An essential app for most people these days, Facebook is the most popular social networking site around. Use this app to make it easier to keep up with your friends, family and colleagues from your Tab, rather than using a browser to go to the site. It also makes updating your statuses, uploading pictures and sharing from other apps or networks easier to do.
2. Instagram – Yet another popular social site, this one relies mainly on sharing thoughts via video or pictures. You can snap a photo on the go right inside Instagram and then edit it to your liking before posting, or even upload a photo from your tablet's files. Share your photos and videos to other popular sites such as Facebook or Foursquare even easier while on the go!

3. The Drawing Pad – An easy app for the kids with its "box of crayons," this app is also great for sketching ideas for business and personal projects, or other fun designs. This one costs $1.99 to purchase at the time of this publication, but is certainly worth checking out.
4. Papyrus – The greatest note-taking app you'll use. It is better than Samsung's note-taking app, and it is free to install!
5. **Kies Aier** – This free app lets you control your Samsung tablet from your TV with no cable required.
6. Story Album – This app helps to arrange photos by date and by location and print them as hardcover books, or print them directly using USB or WI-FI.
7. Flipboard – This is an excellent news aggregator and will get better on Samsung tablets as time wears on.
8. Samsung Smart Camera – This great app can be used to download pictures from your camera or you can use your tablet as a remote, viewfinder, and triggering the camera from a distance.
9. **Xbox IR Remote** – If you've got the popular Xbox game system, this app allows the remote control of your Xbox 360 through its IR port.
10. Kindle – Use this app to access Amazon's eBook store and download your favorite books for reading at your leisure. If you've already purchased a good deal of eBooks from Amazon, you can install the app, log in and start enjoying your books on the Tab 4.
11. Firefox Browser for Android – The popular web browser is optimized for tablets, and offers the ability to sync bookmarks and history with the desktop version. If you use this browser more than Chrome, you'll want to have it with you on the tablet.
12. News360: Personalized News – A great news aggregator, which gets its content from over 10,000 different news sites. It learns from your social accounts and RSS feeds in regard to what you enjoy reading and responds accordingly.
13. **Comics** – Are you or someone in the family into comic books? This app provides a combined comics store and reader app that offers titles from both Marvel Comics and DC comics.
14. Google Earth – The satellite app will give a great indication of everything on earth from overhead. Zoom in for more details of a

specific location, or zoom out for a larger bird's eye perspective. This app can great to use since the tablet has GPS built in.

15. Accuweather – Catch the latest weather forecast so you can get out of its way as it comes to your neighborhood. Use this in addition to the pre-installed weather reading that's on your Tab to get the most complete weather information for your location and other locations of interest.

16. Google Skymap – Hold your tablet in the air during the night and you will get an immediate guide to what all those stars are called, and a great way to find the planets too.

17. OpenTable – A great way to find nearby restaurants, and putting them on the map with a description of the facility and its cuisine.

18. Sketchbook Express **(or Pro)** – AutoDesk's drawing and painting app helps you in your on-screen sketching. Designed for the pros, but easy enough for the rest of us. Easy to sketch on and get good looking results. The Express version is free, or you can opt for Sketchbook Pro for even more features.

19. Magisto Video Editor & Maker – Make stunning videos and edit their content on the fly. Easy to use and very handy. Magisto seamlessly combines your video clips with music and themes, making them as professional as they can be, and it works very quickly, like magic.

20. Floating Browser – The Floating app brings up a browser that opens a web window at just a touch, and lets you use the rest of the screen for other tasks.

21. Evernote – The best note-taking app that has everything that you need to get up to speed in the productivity sphere. Never again will you lose a thought or a note to yourself, as they are simple to store and to fine in a second. Great for those involved in business projects as they collaborate with others and keep track of all the fine details.

22. Dropbox – This is one of the more well-known cloud storage services allowing 2GB of storage and file sharing capabilities, keeping all of your data backed up. There are options to upgrade for more storage space as needed and the interface on the Galaxy Tab 4 is intuitive, making this a great addition to your device.

23. Netflix - If you have a Netflix account, or want to get one, having the app on your tablet will allow you to stream media from anywhere, so you can catch up on your favorite programs or simply catch a movie on a long train ride. You can start the program at home, pick up where you left off on your device and even enjoy it from your computer at work to catch the end. Netflix will stop where you stop, and allow you to pick up where you left off without skipping a beat. You can even start more than one program, and it will hold your spot for you until you are ready to resume.

24. ESPN SportsCenter - This sports app allows you to choose your favorite teams within, and receive real time updates on when they start a game, score or close out a win (or loss). Trades, releases, drafts, and all of the latest sporting information is available not only for the teams you choose within the app, but for all of the latest sporting headlines and scores available worldwide. You can also watch videos and read headlines and story summations in a smooth format, without delving into the entire article. Another great app for those who have access is WatchESPN, which will allow you to log in and watch live streaming videos of sports events from ESPN on your device.

25. Android Device Manager - Apple has "Find my iPhone" and Android – which includes the S5 phone and Tab lineup of tablets, has Android Device Manager. Google's official device tracking service helps you locate any phone or tablet, whether you left it at a friend's house, or it fell under your car seat and remotely erase its data if it is in a compromising place remotely. You can reset the screen lock PIN, and it works with any device associated with your Google account, so you can apply the manager to any of your devices and always know where you stand. This app is perfect for those who own one or multiple Android devices.

10 Great Games for Samsung Tab 4

While apps can help you stay productive, keep up with friends, family or co-workers and do a variety of helpful tasks, games tend to help cure boredom, pass the time, and even sharpen your mental skills. There are thousands of games currently available for Android at the Google Play Store. Here are 10 great games to try, including both free and pay titles that you are sure to enjoy on your device.

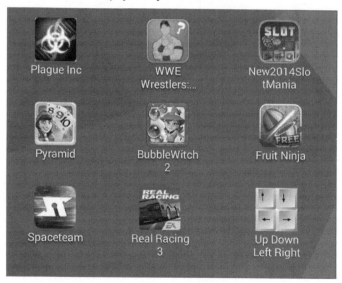

New 2014 Slot Mania

Slot machine games are all the rage, and if you love this Vegas-style game, this is a great one to use to get your slots on. See how far you can drive up your score on the Galaxy Tab 4!

Fruit Ninja

There are two versions of this touch-enabled casual game. One for free, and one for $0.99. The game is simple, as fruit appears on the screen use your touch screen to slash and cut them.

Spaceteam

Spaceteam is a free local multiplayer game that uses Wi-Fi or Bluetooth connectivity to challenge users to work together to save their space ships. It's a great party game, and what's more it's cross-platform, so your iPad owning friends can join in the fun too.

Zippy Zap Space Tap

A fun free game that is similar to Don't Step on the White Tile. In this space-themed version, you're guiding Zippy, a space alien from planet to planet safely in his UFO. As you move to higher levels, the game speeds up. See how high you can push your score by tapping planet to planet!

Plague Inc.

If you're bored with trying to save the world, why not try to destroy it instead. Plague Inc. is a free simulation game where you try to build a virus that will wipe out the world's population. As scientists work to create a cure and governments lock down their countries to slow the spread, your job is to mutate the virus, enabling it to continue to spread, and continue to kill.

Real Racing 3

Real Racing 3 is a free-to play (but with some rather expensive microtransactions) racing game from EA. This game features spectacular visuals for a tablet game, and a good driving model too.

Football Manager Handheld 2014

Football Manager Handheld 2014 is a portable version of the critically acclaimed Football Manager game. It is quite expensive for a tablet game, but it is fully featured, and a very polished handheld title.

WWE Wrestling Superstars Guess

Fans of professional wrestling will want to check out this free game at Google Play. The app provides "cartoonized" images of different pro wrestlers and divas from the WWE for fans and their friends to try to guess. Enter the correct name and move ahead while winning gold coins. Use coins to buy hints for wrestlers you get stumped and see how many you can guess!

Airport Mania: First Fight XP

Airport Mania is a cute and entertaining Diner-Dash like title that is ideal for this time of year. At just $0.99, this game is great value for money. It is a pick-up and play title that is cute enough for kids, but challenging enough (in the later levels) to entertain adults too.

Up Down Left Right

This free Android game is a great brainteaser that challenges you to press the right buttons at the right time. The premise is simple, press up, down, left or right when those arrows appear at the bottom of the screen. The catch - at any moment in time one of the directions is changed, so you may be told to press up, but actually have to press right. It sounds simple, but it's actually fiendishly hard.

Troubleshooting Galaxy Tab 4

Sometimes you will have issues with your Galaxy Tab 4. One of the first things to try for any general trouble is to simply turn off the Galaxy Tab 4 and restart the tablet.

Of course, restarting does not solve every issue with the tablet. Below are some solutions to common issues with this Samsung tablet.

Tablet Won't Turn On

If the Galaxy Tab 4 tablet isn't turning on when you hold the power button down, it could be one of many things. The first step is to try to charging the device and see if it holds a charge. If it doesn't, then the battery inside the tablet might need to be replaced by the manufacturer or an authorized repair service.

There could be a software issue if your device won't turn on, so a reinstallation of the operating system could be another consideration. Consult the manufacturer's website for the product for instructions on this.

The other things to consider if those fixes do not work are that the power button is jammed or broken, or the device has a broken motherboard, both of which can be fixed by the manufacturer.

Tablet Won't Turn Off

If the Galaxy Tab 4 will not turn off, try holding the power button for 8 seconds. If the 8 seconds trick does not work, let the tablet set idle for at least 12 seconds, and then try holding the power button down for 8 seconds again. This should turn the tablet off.

Tablet Won't Charge

This particular issue could be related to a dead battery, or it could also be the result of a faulty charger, or some sort of blockage for the connection port on your tablet. For a dead battery, you'll need to replace it, which you can do yourself, by sending it to the manufacturer or bringing it to an authorized electronics repair shop.

In the case of the faulty charger, try another charger if possible to see if that works (you may need to purchase an OEM version for less money). In the case of the connection port being blocked, you'll need to do your best to carefully clean the port out. Use a thin item carefully such as a paper clip or needle to remove debris. You can also try lightly blowing into the port area to free up any dirt or debris.

Tablet Freezes Issue

If your device is experiencing "freezes" where it appears to lock up suddenly, there are several ways to troubleshoot the issue. First, attempt to power down the device by holding down the power button. Once it's powered off, hold the button to turn the tablet back on. This may resolve the freeze.

Check to see if you've filled up your device's storage capacity next. You can do this by tapping on "Settings" and then going to "General" and "Application manager." You can see at the very bottom of the screen how much memory is currently being used and how much is free. Check both the "Downloaded" and "SD CARD" options at the top of the screen to see if you might need to free up space on your device.

A freeze could also be the result of a recent app you've installed conflicting with the device's operating system or other settings. Try uninstalling the most recent app to see if that solves the issue. You may need to uninstall the last several apps, especially if they were from third party app makers, just to see if it was one of these apps causing the issue.

Next up, try scanning your device with an anti-virus app or program. These can help to detect any sorts of malware, viruses or spyware that possibly have infected your device.

The last resorts to solving this issue are the hard (factory) reset of your device as outlined in the instructions in this guide, having it sent to the manufacturer, or bringing it to an authorized electronics repair service.

Wi-Fi Connectivity Issues

If you are having a Wi-Fi connectivity issue with the Galaxy Tab 4, there are a few considerations before considering the device defective. First and foremost, make sure that the Wi-Fi is switched to on for your Galaxy Tab 4 and that the appropriate network is found. Tap on that network to connect. If the device is failing to connect, try the next solution.

Your modem and wireless router may need to be restarted. Unplug both, then wait a few minutes before plugging them back in. Try to re-connect with the Tab 4, making sure you have the correct network and any passwords needed to log in.

If these solutions don't work, it is probably time to contact your Internet service provider to find out if there's an Internet connectivity issue on their end or with your device. As always, make sure to consult all documentation with your modem, router and Internet provider when trying to troubleshoot Wi-Fi issues.

App Suddenly Stops

What do you do when an app suddenly stops while you are using your Galaxy Tab 4? Sometimes you will receive and error message about the app, and you can select the "Force Close" option, and that will close the app.

If you do not get the error message, you can also manually shut down an app. To manually shut down an app, do the following:

1. Choose the Settings icon from the Apps Menu.
2. Choose Applications.
3. Choose Manage Applications.
4. Choose the application that is suddenly stopping.
5. Touch the Stop button.

If you continue having trouble with the app, you should try uninstalling the app and then reinstalling it. If that does not work, contact the app's developer.

Touchscreen Won't Work

First make sure you are not wearing regular gloves. The only types of gloves you can use with the touchscreen are those that are specially designed to use with touchscreen devices.

If you are not wearing gloves, check the tablet's power. If it is low, simply charge the device, and try the touchscreen again after the tablet has charged up.

If neither of those suggestions fixes the issue, try restarting the Galaxy Tab 4.

Tablet is Hot to Touch

If the tablet is hot to the touch, try turning it off and letting it cool down. Sometimes if it is left in a hot car or has been used for a long amount of time in one setting, the Galaxy Tab 4 may become hot to the touch. While some heat is normal, it should not be too hot to hold. If allowing the tablet to cool down does not fix the issue, the battery may need to be replaced.

Additional Samsung Galaxy Tab 4 Accessories

Purchasing the Samsung Galaxy Tab 4 is an exciting proposition for anyone, but there is more to the purchase than meets the eye. In order to get the most out of this product, you might consider purchasing several additional accessories to maximize its performance. Let's take a look at some of the items to consider for your device.

Protective case and screen protector

Sturdy and built to protect, a protective case and screen protected by Otterbox is essential. Offering three layers of unyielding protection, this is an exceptional solution for securing the Samsung Galaxy Tab 4. The case ensures complete access to all buttons and features while remaining lightweight.

Another consideration for an accessory to purchase along with your case is a screen protector. These are generally sold in packs of 3 and will offer protection from smudges, scratches and dirt on your screen.

Samsung - Gear Fit Fitness Watch with Heart Rate Monitor

For all fitness enthusiasts, this immersive and well-designed heart rate monitor is a modern solution for your fitness requirements. The Samsung Gear Fitness Watch tracks the user's calories burned, heart rate, distance covered, number of steps taken, and speed. The monitor also features an excellent interface (Bluetooth 4.0 + LE) designed to enable users to sync results with the Samsung Galaxy Tab 4. You'll also want to add the apps Gear Manager and/or Gear Fit Manager on your tablet to work along with the health device.

Bluetooth headset

Hands-free talking has never been easier than using a quality Bluetooth headset. This premium device utilizes resonating sound (HD) and revolutionary audio enhancements to produce a crystal clear experience for yourself and the individual on the line. With comprehensive multi-connectivity functionality, it is easier to switch without trouble between multiple lines (i.e. cell phone and work line).

Among the headsets to take a look at for your Galaxy Tab 4 is the Sennheiser VMX 200 model.

Bluetooth Speaker

Tiny enough to bring with you anywhere, yet big enough to deliver immersive, crystal clear sound, the Bluetooth speaker fits nicely into your palm. High-quality wireless technology enable you to stream sound from your device. The product comes with a charging cradle to enable the user to take it anywhere, and can be customized with a range of uniquely colored covers. The JBL Flip is a highly rated Bluetooth speaker that also functions great with many smartphones to play music and also accept phone calls via the speaker.

Bluetooth Keyboard

Convert your Samsung Galaxy Tab 4 into an immersive machine with this modernized keyboard packed with Bluetooth connectivity. The keyboard pairs wirelessly with your device and the protective case turns into a firm, stylish stand that can support the device in either portrait or landscape mode.

Use a compact, low profile keyboard that is portable and comfortable to type on. Uniquely styled keys ensure the user is easily able to type and navigate on the device. Equipped with specialized media controls to pause, play, volume up, and volume down. The Bluetooth keyboard is designed to have a wireless range of up to thirty feet. There are a variety of Bluetooth keyboards on the market that will meet style and functionality needs, with some even doubling as a protective case for your device.

Wired stereo headphones or earbuds

Want to listen to music? It is important to use quality wired headphones such as Beats by Dre, Sennheiser or Bose to get crystal clear sound quality.

A lightweight, on-ear pair of headphones, the Beats Solo is well sized and compact enough to easily tuck away into your bag or purse. These headphones encompass the smooth, powerful sound that has ensured Beats by Dr. Dre is one of the most iconic brands in the world. Bose makes excellent styles including noise-filtering types perfect for travel, while Sennheiser has some great earbuds for a variety of purposes.

MicroSD storage card (Kingston MicroSDHC Class 32GB)

The Tab 4 tablets will accept additional storage using either a MicroSD or MicroSDHC card. The Tab 4 7-inch will accept as large as a 32GB card, while the 8 and 10.1 tablets will allow up to 64GB cards. You can easily place all of your most important content and files in a supremely designed Kingston MicroSDHC storage card. Compatible with your Samsung Galaxy Tab 4, this brand of MicroSDHC card is a powerhouse when it comes to storing content ranging from games, movies, music, and much more. Want to get more out of this accessory? Utilize the adaptor that comes with many different brands of these cards to change it into an all-out SDHC card.

Where to get More Help with Galaxy Tab 4

If you have mobile service for your Galaxy Tab 4, the first place to ask for help is with your service provider. If you just have the Wi-Fi version of the tablet, you can contact Samsung's customer support at 1-800-726-7864.

Below are other forums where you may find help with your tablet.

Galaxy Tab at Android forum
<http://androidforums.com/galaxy-tab-tips-tricks/>

Galaxy Tab 4 Forum on XDA <http://forum.xda-developers.com/tab-4>

The Galaxy Tab Forum
<http://www.thegalaxytabforum.com/index.php?/forum/14-galaxy-tab-original-7-inch-user-help-solutions/>

Samsung Support
<http://www.samsung.com/us/support/>

Conclusion

The Galaxy Tab 4 tablets are a versatile group of tablets from Samsung's technology products. The 7, 8 and 10.1 models of the Tab 4 are great mid-range tablets that can do the majority of what you need to them do. Additional apps and accessories will help you get even more from the device. Overall, the Tab 4 offers good value for the money spent. There's always new features being developed and added by Samsung, so look for future updates.

As you can see, the tablet will offer plenty of entertainment and productivity to meet everyday needs. Simply spend a bit of time getting to know your tablet, and learn how to use it to its fullest capacity. If you do this, you will get the most out of your Galaxy Tab 4, and will surely enjoy owning this latest innovation from Samsung.

More Books by Shelby Johnson

Samsung Galaxy S5 User Manual: Tips & Tricks Guide for Your Phone!

Amazon Fire TV User Manual: Guide to Unleash Your Streaming Media Device

Apple TV User's Guide: Streaming Media Manual with Tips & Tricks

iPad Mini User's Guide: Simple Tips and Tricks to Unleash the Power of your Tablet!

iPhone 5 (5C & 5S) User's Manual: Tips and Tricks to Unleash the Power of Your Smartphone! (includes iOS 7)

Kindle Fire HDX & HD User's Guide Book: Unleash the Power of Your Tablet!

Facebook for Beginners: Navigating the Social Network

Kindle Paperwhite User's Manual: Guide to Enjoying your E-reader!

How to Get Rid of Cable TV & Save Money: Watch Digital TV & Live Stream Online Media

Chromecast User Manual: Guide to Stream to Your TV (w/Extra Tips & Tricks!)

Google Nexus 7 User's Manual: Tablet Guide Book with Tips & Tricks!

Roku User Manual Guide: Private Channels List, Tips & Tricks

Chromebook User Manual: Guide for Chrome OS Apps, Tips & Tricks!

Printed in Great Britain
by Amazon.co.uk, Ltd.,
Marston Gate.